We have the pleasure of sul

Science's Strangest]
$ 13.00 pb
Tom Quinn

U.S. Publication date October 15, 2005
A clipping of your review
would be appreciated
800-423-4525

Chrysalis
Distributed by
Trafalgar Square
No. Pomfret, Vermont 05053
www.trafalgarsquarebooks.com

B...
Brid...
Cinema...
Cricket's ...
Fishing's S...
Football's Stran...
Golf's Stranges...
Horse-Racing's Stran...
The Law's Strangest ...
Medicine's Strangest C...
Motor-racing's Strangest Races
The Olympics' Strangest Moments
Politics' Strangest Characters
Railway's Strangest Journeys
Royalty's Strangest Characters
Rugby's Strangest Matches
Sailing's Strangest Moments
Shooting's Strangest Days
Tennis's Strangest Matches